Julie K. Lundgren

A Crabtree Seedlings Book

# TABLE OF CONTENTS

# LIVING IN WATER

Fish come in many shapes, colors, and sizes. They have backbones, **gills** and fins, and are **ectotherms**. As water temperatures rise and fall, their bodies heat and cool to match.

## Creepy or Cool?

*The opah, a deep ocean fish, is the only true warm-blooded fish.*

*A whale shark is the largest fish in the sea. It can grow up to 40 feet (12 meters) or more.*

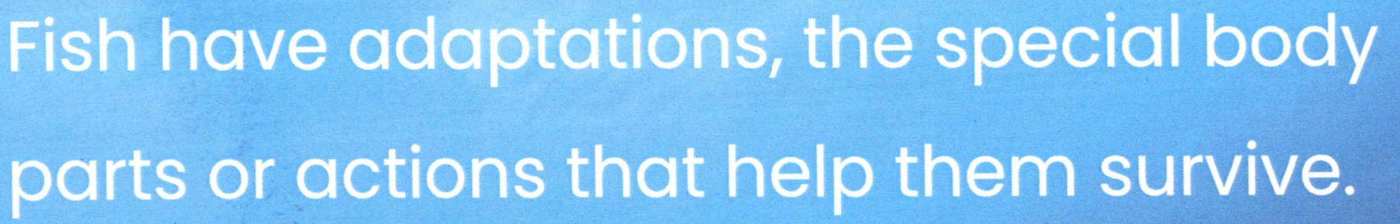

Fish have adaptations, the special body parts or actions that help them survive.

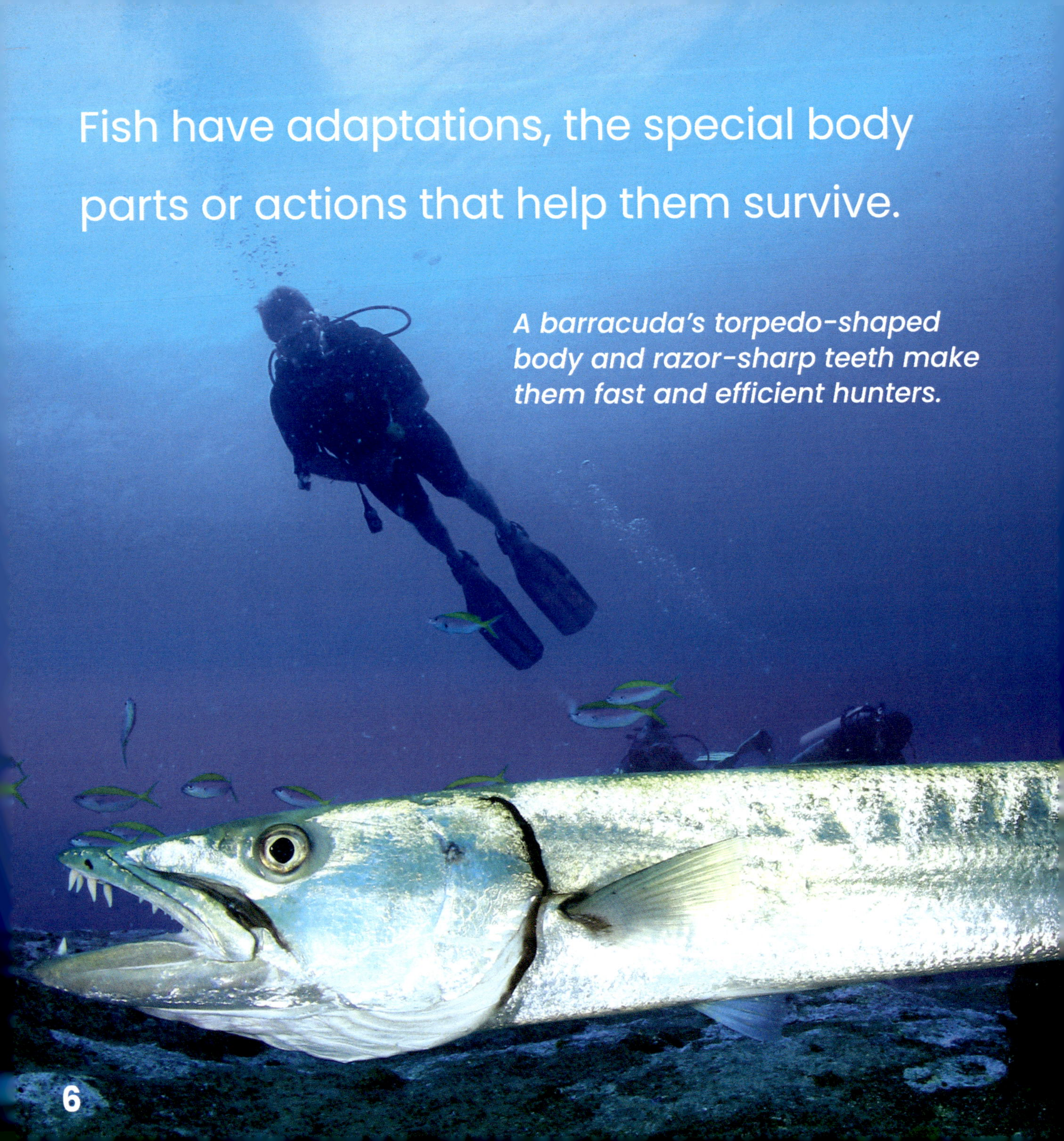

*A barracuda's torpedo-shaped body and razor-sharp teeth make them fast and efficient hunters.*

## CREEPY OR COOL?

*Mudskippers have special body parts that allow them to live on land. They fill their gill chambers with water for breathing. They use their fins to crawl. They use their mouths like shovels to dig wet holes to hide in.*

Fish live in saltwater and freshwater **habitats**. These habitats can be cold or warm, light or dark, deep or shallow.

*Alligator gars, a kind of river fish, can grow 10 feet (3 meters) long. They have a snout shaped like a crocodile's and rows of sharp teeth.*

*Green moray eels quietly hide in coral reefs, awaiting **prey**.*

attacks through actions.

*Fish may find safety and better hunting success in large groups, called schools.*

*Pufferfish take in water to swell their bodies like balloons, making them hard to swallow.*

Other fish defend themselves through special body adaptations.

*When attacked, hagfish make thick slime for a slippery escape.*

*Bright colors remind **predators** not to eat poisonous red lionfish.*

# FOOD FINDERS

Fish must eat and avoid being eaten. Some use **camouflage** to hide from predators or from prey.

*Oyster toadfish can change color to match their environment.*

*Leafy sea dragons sway among sea grasses.*

Fish have special body parts to capture their favorite foods.

*Catfish have thick, spongy whiskers covered in taste buds for finding food in dark waters.*

## CREEPY OR COOL?

*Sea lamprey have mouths with circular rows of teeth, which they use to attach to other fish. They suck the blood and juices from the fish.*

**Carnivores** find and capture prey. They may use slashing teeth, glowing lures, or even electric shock.

## CREEPY OR COOL?

*Electric eels shock prey to death. EEK!*

*Hairy frogfish are covered in hair-like spines. They attract prey by waving a special body part that looks like a worm.*

Scavengers eat the dead and dying, and other nibbles that float or swim by.

*Remoras swim next to sharks and other large fish and eat their leftovers.*

*Lingcods lay between 60,000 and 500,000 eggs at a time.*

# SMALL FRY, PLEASE

Female fish lay eggs and males fertilize them. Eggs may hatch in days or months, depending on the kind of fish.

Young fish, called fry, grow into adults. A fish is an adult when it can **reproduce**. Only a few eggs survive to become adult fish.

*Eggs and fry make tasty meals for birds and other fish.*

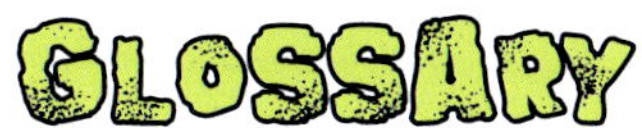


**camouflage** (KAM-uh-flahzh): Colors or patterns that blend with the surroundings to help animals hide

**carnivores** (KAR-nuh-vorz): Animals that eat other animals

**ectotherms** (EK-toh-thermz): Animals whose body temperatures change with the environment

**gills** (GILZ): A special body part of fish and amphibians that help them breathe underwater

**habitats** (HAB-uh-tats): Places animals naturally live

**predators** (PRED-uh-turz): Animals that hunt and eat other animals

**prey** (PRAY): Animals hunted by other animals for food

**reproduce** (ree-pruh-DOOSS): Make more of something

## INDEX

# School-to-Home Support for Caregivers and Teachers

This book helps children grow by letting them practice reading. Here are a few guiding questions to help the reader build his or her comprehension skills. Possible answers appear here in red.

## Before Reading

- **What do I think this book is about?** *I think this book is about Creepy and cool fish. I think this book is about where different fish live.*
- **What do I want to learn about this topic?** *I want to learn how fish breathe under water. I want to learn if fish can see in the darkest parts of the sea.*

## During Reading

- **I wonder why...** *I wonder why fish come in so many different shapes. I wonder why there are poisonous fish and where do they live.*
- **What have I learned so far?** *I have learned that the whale shark is the largest fish in the sea, it can grow up to 40 feet (12 meters). I have learned that electric eels shock their prey to death.*

## After Reading

- **What details did I learn about this topic?** *I have learned that fish live in saltwater and freshwater habitats. I have learned that pufferfish take in water to swell their bodies like balloons, making them hard to swallow by larger fish.*
- **Read the book again and look for the glossary words.** *I see the word **ectotherms** on page 4, and the word **gills** on page 4. The other glossary words are found on page 23.*

---


**Library and Archives Canada Cataloguing in Publication**

Title: Fish / Julie K. Lundgren
Names: Lundgren, Julie K., author.
Description: Series statement: Creepy but cool | "A Crabtree seedlings book". | Includes index. | Previously published in electronic format by Blue Door Education in 2019.
Identifiers: Canadiana (print) 20210201762 | Canadiana (ebook) 20210201770 | ISBN 9781427161666 (hardcover) | ISBN 9781427161789 (softcover) | ISBN 9781427161901 (HTML) | ISBN 9781427162021 (EPUB) | ISBN 9781427162144 (read-along ebook)
Subjects: LCSH: Fishes—Juvenile literature.
Classification: LCC QL617.2 .L86 2022 | DDC j597—dc23

**Library of Congress Cataloging-in-Publication Data**

Names: Lundgren, Julie K., author.
Title: Fish / Julie K. Lundgren.
Description: New York : Crabtree Publishing, [2022] | Series: Creepy but cool - a Crabtree seedlings book | Includes index.
Identifiers: LCCN 2021018421 (print) | LCCN 2021018422 (ebook) | ISBN 9781427161666 (hardcover) | ISBN 9781427161789 (paperback) | ISBN 9781427161901 (ebook) | ISBN 9781427162021 (epub) | ISBN 9781427162144
Subjects: LCSH: Fishes--Juvenile literature.
Classification: LCC QL617.2 .L88 2022 (print) | LCC QL617.2 (ebook) | DDC 597--dc23
LC record available at https://lccn.loc.gov/2021018421
LC ebook record available at https://lccn.loc.gov/2021018422

**Crabtree Publishing Company**
www.crabtreebooks.com 1-800-387-7650
Print book version produced jointly with Blue Door Education in 2022

Written by Julie K. Lundgren
Print coordinator: Katherine Berti
Printed in Canada/042023/CPC20230420





Photo credits:
Cover ©Kletr; Pages 4-5 © VisionDive; Page 5 © NOAA Fisheries; pages 6-7 ©Rich Carey, page 7 inset photo © twospeeds; Pages 8-9 ©Andrea Izzotti, page 9 inset photo ©Miguel Aleixo; Page 10 ©Leonardo Gonzalez, Page 11 ©Moize nicolas; page 12 ©NOAA photo library https://creativecommons.org/licenses/by/2.0/, page 13 ©Gilmanshin; pages 14-15 © Kris Wiktor, page 14 inset photo Joe Quinn; pages 16-17 © Kletr, page 17 inset photo © Drow_male https://creativecommons.org/licenses/by-sa/3.0/deed.en; pages 18-19 © Tamil Selvam, page 18 inset photo ©Steven G. Johnson https://creativecommons.org/licenses/by-sa/3.0/deed.en; page 20 ©VisionDive, page 21 © NatureDiver; page 22 ©Kletr,
All photos from Shutterstock.com unless otherwise stated.

**Published in the United States**
**Crabtree Publishing**
347 Fifth Ave.
Suite 1402-145
New York, NY 10016

**Published in Canada**
**Crabtree Publishing**
616 Welland Ave.
St. Catharines, Ontario
L2M 5V6